# Mental Math
# and
# Estimation

*Don Miller*

*Cuisenaire Company of America*
*White Plains, NY*

Cover and text design: Nancy R. Naft

# Table of Contents

# Introduction

Recent recommendations by the National Council of Teachers of Mathematics list mental math and computational estimation as skills that every student should develop. The fact that many everyday situations call for an estimate leaves little doubt that some degree of proficiency in making guesses should be expected of students at all ability levels. Students should be able to make a quick mental estimate to decide whether a written or calculator answer is reasonable.

Also, for many situations, an estimate is all that is really needed. It appears, however, that little attention is being given to these important skills since national assessment tests continue to show that an alarming number of students are deficient in these areas. Research shows that students can develop mental and estimation skills, but only when they are provided with appropiate activities and instruction on a regular and sustaining basis.

The purpose of this book is to provide a ready source of activities designed to help students improve their mental and estimation skills. The activities were designed to make a calculator an integral part of the ideas being explored. Besides providing immediate feedback while completing many of the exercises, a calculator can serve as a valuable aid in developing mental math skills.

In using these materials it should be kept in mind that they were designed to supplement the existing mathematics curriculum. They are not meant to be taught as a unit on mental math and estimation. The ideas should be presented on a regular basis during the entire school year. The activities need not be used in the order presented. For use in the classroom, teachers should consider student backgrounds in selecting activities that may be appropiate either to introduce new strategies or to reinforce those previously discussed.

Many pages were designed to serve as transparency masters. The purpose was twofold–to eliminate the need to duplicate the activity for each student and to enhance class discussion. The Test Yourself and Estimation Check activities were designed to help students evaluate their progress both in finding exact answers mentally and in making estimates.

For every activity, teachers should encourage students to discuss both their methods and their answers first in small groups, then in whole class discussions. Students should be encouraged to look for patterns, to explain their thinking, and to use calculators to check results. "Why do you think that?" should be the teacher's response to every answer, reasonable or not.

# Mental Math

Find exact answers mentally.

**1.** 63 + 89 + 47

**2.** 368 + 276

**3.** 642 − 298

**4.** 1.5 x 4400

**5.** $13\overline{)3900}$

**6.** $20.00 − $8.89

**7.** 6 x 580

**8.** 5/8 of 640

**9.** 25% of $3200

**10.** 18 x 450

**11.** 900 − 3 x 120

**12.** 2.5% of $800

**13.** $2.5\overline{)750}$

**14.** 23 x 99

*Mental Math and Estimation*

# Estimation

Find approximate answers mentally.

**1.** $1.89 + $4.99 + $3.19 + $7.88 + $0.79

**2.** 67 x 312

**3.** $15.6 \overline{)469.2}$

**4.** 78.9 x 82 − 879

**5.** 24% of $269

**6.** $5.78 each
Cost for 2 dz?

**7.** 394 miles
15.6 gallons
mpg = ?

**8.** Square
Area = 4761 cm²
Perimeter = ?

**9.** Mel's cafe: $17.96
15% tip = ?

# Mental Addition

| | | |
|---|---|---|
| **67 + 85** | Mark's method: | 60   140   147   152 |
| | Gina's method: | 140   152 |
| **56 + 87 + 78** | Michael's method: | 50   130   20 →206   213   221 |
| | Lorie's method: | 200   221 |
| **374 + 568** | Terry's method: | 800   930   932 |

Use the above ideas to find the answers mentally.  The * number is the sum of the correct answers.

1. 86 + 79          37 + 68          49 + 76          54 + 38          * 487

2. 29 + 68          56 + 37          84 + 77          38 + 75          * 464

3. 47 + 68 + 35          38 + 69 + 74          77 + 66 + 43          * 517

4. 56 + 75 + 87          29 + 76 + 52          47 + 76 + 84          * 582

5. 44 + 59 + 68          74 + 47 + 56          38 + 87 + 27          * 500

6. 375 + 226          585 + 116          425 + 276          * 2003

# Mental Math

Use your mental math skills to find the missing numbers.  The * number is the sum of the correct answers.

---

_____ + \$3.85 = \$10 ⟶ + \$0.15 (\$4) + \$6 (\$10) ⟶ <u>\$6.15</u>

---

**1a.** _____ + \$6.25 = \$10

_____ + \$4.95 = \$20

_____ + \$1.66 = \$5

_____ + \$3.89 = \$10

_____ + \$2.65 = \$5

\* \$30.60

**1b.** _____ + \$13.75 = \$20

_____ + \$8.88 = \$20

_____ + \$6.90 = \$10

_____ + \$24.50 = \$50

_____ + \$12.95 = \$20

\* \$53.02

---

_____ + 2.4 = 20 ⟶ + 0.6 (3) + 17 (20) ⟶ <u>17.6</u>

---

**2a.** _____ + 12.5 = 20

_____ + 77.9 = 90

_____ + 39.5 = 50

_____ + 8.3 = 40

_____ + 28.8 = 50

\* 83

**2b.** _____ + 9.8 = 60

_____ + 16.4 = 40

_____ + 22.8 = 30

_____ + 43.5 = 70

_____ + 33.9 = 80

\* 153.6

---

# Mental Math Challenges

Use your mental math skills to find the missing numbers. Check with a calculator after recording your mental answers.

Example:

|  | +90 | +250 | +700 | +670 |
|---|---|---|---|---|
| 60 | 150 | 400 | 1100 | 1770 |

|  | +48 | +430 | +97 | +269 |
|---|---|---|---|---|
| **1.** 27 | | | | |

|  | +74 | +456 | +67 | +257 |
|---|---|---|---|---|
| **2.** 46 | | | | |

|  | +208 | +65 | +207 | +244 |
|---|---|---|---|---|
| **3.** 46 | | | | |

|  | +96 | +347 | +88 | +144 |
|---|---|---|---|---|
| **4.** 76 | | | | |

|  | +69 | +75 | +306 | +299 |
|---|---|---|---|---|
| **5.** 74 | | | | |

|  | +87 | +319 | +78 | +198 |
|---|---|---|---|---|
| **6.** 138 | | | | |

*Mental Math and Estimation*

# Addition Estimation (Your Choice)

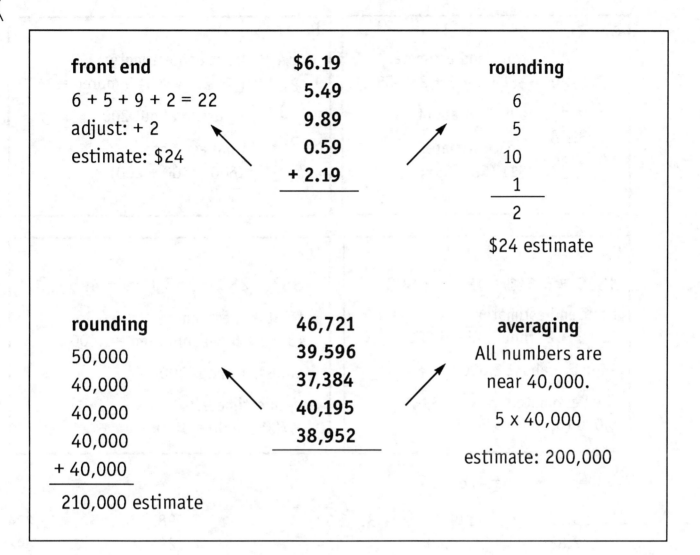

**front end**

6 + 5 + 9 + 2 = 22

adjust: + 2

estimate: $24

$6.19
5.49
9.89
0.59
+ 2.19

**rounding**

6
5
10
1
‾‾‾
2

$24 estimate

**rounding**

50,000
40,000
40,000
40,000
+ 40,000
‾‾‾‾‾‾
210,000 estimate

46,721
39,596
37,384
40,195
38,952

**averaging**

All numbers are
near 40,000.

5 x 40,000

estimate: 200,000

Try these—estimates only.

| **1.** | **2.** | **3.** | **4.** |
|---|---|---|---|
| 706 | $4.88 | 368 | $3.21 |
| 691 | 0.29 | 2918 | 1.76 |
| 687 | 9.99 | 75 | 0.98 |
| 729 | 3.19 | 421 | 2.89 |
| 801 | 1.49 | 5039 | 1.19 |
| + 649 | + 0.59 | + 3276 | + 19.09 |

# Addition Estimation (Front End)

**A.**

$1.79
2.29
0.97
3.38
+ 2.89

front end estimate:
$(1 + 2 + 3 + 2) = $8

adjust: + about $3

final estimate:
$11 ($8 + $3)

**B.**

349
227
439
506
+ 89

front end estimate:
(3 + 2 + 4 + 5) hundred = 1400

adjust: + about 200

final estimate:
1600 (1400 + 200)

**C.**

1369 + 3,641 + 187 + 2,898 = ?

front end estimate:
(1 + 3 + 2) thousand = 6,000

adjust: + about 2,000

final estimate:
8,000 (6,000 + 2,000)

**D.**

387 + 239 + 568 + 19 + 495 = ?

front end estimate:
(3 + 2 + 5 + 4) hundred = 1,400

adjust: + about 300

final estimate:
1,700 (1,400 = 300)

Try these—estimates only.

| **1.** | **2.** | **3.** | **4.** | **5.** |
|---|---|---|---|---|
| $3.19 | $19.99 | 398 | 3,098 | 19,234 |
| 2.88 | 2.49 | 129 | 29 | 698 |
| 0.29 | 0.25 | 15 | 2,798 | 6,108 |
| 4.25 | 3.39 | 439 | 876 | 39 |
| + 6.39 | + 31.88 | + 79 | + 7,319 | + 8,875 |

**6.** $3.95 + $0.29 + $4.89 + $6.88 + $0.78 + $1.19 + $8.05 ⟶ _____

**7.** 345 + 19 + 819 + 699 + 89 + 207 + 402 ⟶ _____

**8.** 8,199 + 698 + 17 + 12, 775 + 1,899 + 5,359 ⟶ _____

*Mental Math and Estimation*

# Mental Subtraction

Kim's method

$74 - 38$

$+2$   $+2$

$76 - 40 = 36$

$74 - 38$

Randy's method

$74 - 40 = 34$

$34 + 2 = 36$

Kim's method

$425 - 97$

$+3$     $+3$

$428 - 100 = 328$

$425 - 97$

Randy's method

$425 - 100 = 325$

$325 + 3 = 328$

Use the above ideas to find the answers mentally.  The * number is the sum of the correct answers.

1. $73 - 45$     $84 - 57$     $91 - 39$     $83 - 26$     * 164

2. $92 - 57$     $314 - 97$     $63 - 17$     $71 - 18$     * 351

3. $741 - 96$     $503 - 75$     $825 - 89$     $362 - 89$     * 2082

4. $423 - 198$     $602 - 85$     $721 - 297$     $514 - 75$     * 1605

# Mental Math Challenges

Use your mental math skills to find the missing numbers. Check with a calculator after recording your mental answers.

Example:

| | −400 | | −79 | | −199 | | −50 | |
|---|---|---|---|---|---|---|---|---|
| 1300 | | 900 | | 821 | | 622 | | 572 |

| | | −175 | | −199 | | −95 | | −102 | |
|---|---|---|---|---|---|---|---|---|---|
| **1.** | 800 | | | | | | | | |

| | | −650 | | −160 | | −260 | | −88 | |
|---|---|---|---|---|---|---|---|---|---|
| **2.** | 1400 | | | | | | | | |

| | | −170 | | −290 | | −97 | | −105 | |
|---|---|---|---|---|---|---|---|---|---|
| **3.** | 780 | | | | | | | | |

| | | −1990 | | −800 | | −220 | | −99 | |
|---|---|---|---|---|---|---|---|---|---|
| **4.** | 5000 | | | | | | | | |

| | | −975 | | −295 | | −260 | | −91 | |
|---|---|---|---|---|---|---|---|---|---|
| **5.** | 1850 | | | | | | | | |

| | | −87 | | −98 | | −299 | | −89 | |
|---|---|---|---|---|---|---|---|---|---|
| **6.** | 610 | | | | | | | | |

# Mental Math Challenge

Use your mental math skills to find the missing numbers. The number in each square must be the sum of the two adjacent circled numbers.

Example:

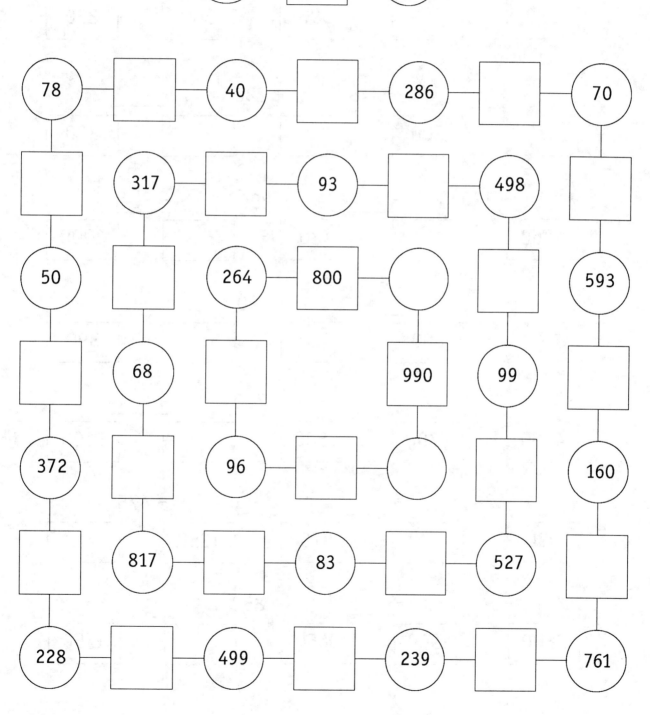

# Mental Math Challenges

Use your mental math skills to find the missing numbers.  Check with a calculator after recording your mental answers.

Example:

|  | + 170 |  | − 360 |  | + 410 |  | − 570 |  |
|---|---|---|---|---|---|---|---|---|
| 580 |  | 750 |  | 390 |  | 800 |  | 230 |

**1.**

|  | + ___ |  | − 348 |  | + 230 |  | − ___ |  |
|---|---|---|---|---|---|---|---|---|
| 375 |  | 800 |  | ☐ |  | ☐ |  | 300 |

**2.**

|  | + 770 |  | + ___ |  | − 290 |  | + ___ |  |
|---|---|---|---|---|---|---|---|---|
| 1360 |  | ☐ |  | 980 |  | ☐ |  | 1000 |

**3.**

|  | + ___ |  | − ___ |  | + 290 |  | − ___ |  |
|---|---|---|---|---|---|---|---|---|
| 670 |  | 900 |  | 420 |  | ☐ |  | 390 |

**4.**

|  | − ___ |  | − ___ |  | + 700 |  | + ___ |  |
|---|---|---|---|---|---|---|---|---|
| 4500 |  | 3700 |  | 1800 |  | ☐ |  | 5100 |

**5.**

|  | − 280 |  | + ___ |  | + ___ |  | − 475 |  |
|---|---|---|---|---|---|---|---|---|
| 930 |  | ☐ |  | 875 |  | 1250 |  | ☐ |

**6.**

|  | − ___ |  | + ___ |  | + 850 |  | − ___ |  |
|---|---|---|---|---|---|---|---|---|
| 2800 |  | 1900 |  | 2650 |  | ☐ |  | 1750 |

# Subtraction Estimation

|  | | | |
|---|---|---|---|
| 84261 | 8**** | 84*** | 86*** |
| −38475 | −3**** | 38*** | −40*** |
| | 50000 | | 46000 |
| | *rough estimate* | | *better estimate* |

| | | | |
|---|---|---|---|
| 7209 | 7*** | 72** | 75** |
| −3741 | −3*** | −37** | −40** |
| | 4000 | | 3500 |
| | *rough estimate* | | *better estimate* |

Try these—estimates only.

**1.**   5274
         −2768

**2.**   81304
         −36548

**3.**   65253
         −28475

**4.**   9034
         −3977

**5.**   71236
         −43482

**6.**   43062
         −15273

**7.**   40346
         −27487

**8.**   93005
         −66328

**9.**   63311
         −32978

# Mental Products

A.  6 x 700 ——→ 6 x 7 = 42 ——→ 6 x 700 = 4200

B.  30 x 250 ——→ 3 x 25 = 75 ——→ 30 x 250 = 7500

C.  8 x 47 ——→ 8 x (40 + 7) = 320 + 56 = 376

D.  80 x 47 ——→ (8 x 47) x 10 = 376 x 10 = 3760

E.  20 x 150 ——→ (2 x 15) x 100 = 30 x 100 = 3000

Use the above ideas to find the answers mentally.  The * number is the sum of the correct answers.

1.  7 x 900        40 x 12        35 x 20        5 x 4 x 60        *8680

2.  25 x 20        2 x 230        40 x 15        8 x 5 x 9        *1920

3.  6 x 57        8 x 36        6 x 68        30 x 83        *3528

4.  40 x 220        20 x 35        50 x 44        7 x 52        *12064

# Mental Math Challenge

Use your mental math skills to find the missing numbers. The number in each square must be the sum of the two adjacent circled numbers.

Example:

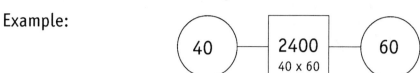

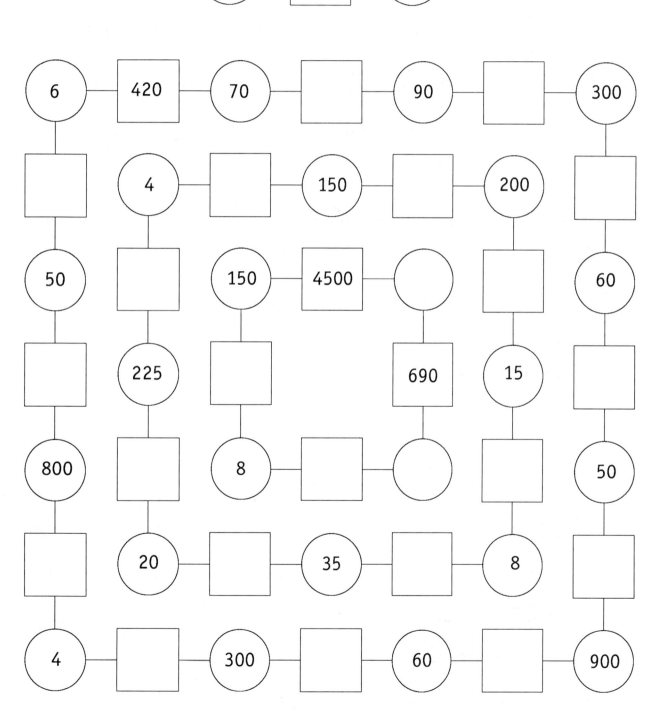

# Estimating Products (Rounding)

> **A.** 43 x 54 → 40 x 50 = 2000   low (both factors rounded down)
>
> **B.** 57 x 73 → 60 x 70 = 4200
>
> **C.** 68 x 29 → 70 x 30 = 2100   high (both factors rounded up)
>
> **D.** 88 x 99 → 88 x 100 = 8800   high (one factor rounded up)
>
> **E.** 21 x 28 → 20 x 28 = 560   low (one factor rounded down)

Try these—estimates only. Indicate when you are sure that an estimate is high (greater than the exact answer) or low (less than the exact answer).

**1 a.** 48 x 27 → _____ x _____ = _____

**b.** 63 x 74 → _____ x _____ = _____

**c.** 58 x 32 → _____ x _____ = _____

**d.** 71 x 72 → _____ x _____ = _____

**e.** 31 x 87 → _____ x _____ = _____

**2 a.** 57 x 78 → _____      **b.** 37 x 66 → _____

**c.** 82 x 89 → _____      **d.** 60 x 91 → _____

**e.** 58 x 71 → _____      **f.** 65 x 45 → _____

**g.** 29 x 218 → _____      **h.** 397 x 52 → _____

**i.** 98 x 59 → _____      **j.** 32 x 148 → _____

# Estimating Products (Front End)

6 x 4,531 ⟶ 6 x 4,000 = 24,000

adjust: + about 3,000 (6 x 500)

estimate: 27,000 (24,000 + 3,000)

$7 x 638 ⟶ $7 x 600 = $4,200

adjust: + about $280 ($7 x 40)

estimate: $4,480 ($4,200 + $280)

82,324 x 7 ⟶ 80,000 x 7 = 560,000

adjust: + about 14,000 (2,000 x 7)

estimate: 574,000 (560,000 + 14,000)

Try these—estimates only.

**1.** 6 x 7,613        **2.** 7 x 448        **3.** 4,286 x 8

**4.** $8 x 4,716        **5.** 8,722 x 5        **6.** 7 x $8.31

**7.** 6 x 46,293        **8.** 6 x 7,528        **9.** 60 x $8.49

# Missing Factor Challenges

Use your mental math skills to guess the missing factors, M and N.  Check each guess with a calculator.  Score 3 points if your first guess is correct and 1 point if it takes two tries to find the correct factors.

**1.** M x N = 814

M = ?  37  67  87
N = ?  52  22  12

**2.** M x N = 2736

M = ?  22  52  72
N = ?  38  68  88

**3.** M x N = 2772

M = ?  33  53  83
N = ?  84  64  24

**4.** M x N = 2146

M = ?  29  69  89
N = ?  44  54  74

**5.** M x N = 1363

M = ?  67  47  27
N = ?  29  49  89

**6.** M x N = 2064

M = ?  28  48  98
N = ?  13  43  63

My score _____

# Three-in-a-Line Estimation

1. Select two of the numbers shown next to the playing board.
2. Write the numbers and then find the product with a calculator. If the answer is on the playing board, mark it with an X.
3. Your goal is to get three Xs in a line (diagonal, vertical, or horizontal) in five or less tries.
4. Score as follows:   3 tries → 5 points/4 tries → 3 points/5 tries → 1 point

**A.**

| | | |
|---|---|---|
| 5119 | 1739 | 799 |
| 629 | 3619 | 2479 |
| 1309 | 1139 | 3149 |

17
37
47
67
77

1. _____ x _____ = _____
2. _____ x _____ = _____
3. _____ x _____ = _____
4. _____ x _____ = _____
5. _____ x _____ = _____
My score _____

**B.**

| | | |
|---|---|---|
| 949 | 3139 | 799 |
| 6789 | 1209 | 3869 |
| 2279 | 559 | 3999 |

13
43
53
73
93

1. _____ x _____ = _____
2. _____ x _____ = _____
3. _____ x _____ = _____
4. _____ x _____ = _____
5. _____ x _____ = _____
My score _____

**C.**

| | | |
|---|---|---|
| 2275 | 1275 | 1925 |
| 3575 | 4675 | 825 |
| 975 | 2975 | 5525 |

15
35
55
65
85

1. _____ x _____ = _____
2. _____ x _____ = _____
3. _____ x _____ = _____
4. _____ x _____ = _____
5. _____ x _____ = _____
My score _____

# Missing Number Challenges

Use a calculator to complete the table.

| x | 1.5 | | 9.8 | |
|---|---|---|---|---|
| 82 | | 2788 | | |
| | | 221 | | |
| | 675 | | | 31725 |

Use the completed table and your mental math skills to guess the *exact* answer. Record your mental answer, then check with a calculator.  Score 1 point for each correct mental answer.

1.  0.82 x 705 =  _____        2.  221 ÷ 3.4 =  _____

3.  45 x 0.98 =  _____        4.  650 x 7.05 =  _____

5.  44.1 ÷ 0.45 =  _____        6.  1530 ÷ 34 =  _____

7.  4.5 x 340 =  _____        8.  97.5 ÷ 0.65 =  _____

9.  6.37 ÷ 0.65 =  _____        10.  57.81 ÷ 0.82 =  _____

11.  9800 x 0.82 =  _____        12.  3.4 x 6500 =  _____

13.  980 x 6.5 =  _____        14.  45.825 ÷ 0.65 =  _____

15.  45000 x 0.034 =  _____        16.  63700 ÷ 980 =  _____

17.  0.675 ÷ 1.5 =  _____        18.  22.1 ÷ 0.34 =  _____

19.  15000 x 0.045 =  _____        20.  0.098 x 820 =  _____

My score _____

*Mental Math and Estimation*

# Analyzing Multiplication

```
        638              497
      x 497            x 638
      ------           ------
       4466             3976
      57420            14910
     255200           298200
     -------          -------
     317086           317086
```

Use the completed problems above and your estimation skills to find the missing numbers mentally. Record your mental answers, then check with a calculator. Score 1 point for each correct estimate.

**1.** 8 x 497 = _____

**2.** 4 x 638 = _____

**3.** 638 x _____ = 574.2

**4.** 38 x 497 = _____

**5.** 63.8 x 4 = _____

**6.** 255.2 = _____ x 63.8

**7.** 6.38 x 49.7 = _____

**8.** 60 x 4.97 = _____

**9.** 97 x 638 = _____

**10.** 14.91 ÷ 4.97 = _____

**11.** 49.7 x 80 = _____

**12.** 63.8 x 4.97 = _____

**13.** 638 x 97 = _____

**14.** 29.82 = 4.97 x _____

My score _____

# Mental Division

| | | | |
|---|---|---|---|
| **A.** | $42 \div 7 = 6$ | $420 \div 7 = 60$ | $4,200 \div 70 = 60$ |
| **B.** | $36 \div 4 = 9$ | $3,600 \div 4 = 900$ | $36,000 \div 90 = 400$ |
| **C.** | $4,800 \div 60 \longrightarrow 48 \div 6 = 8 \longrightarrow 4,800 \div 60 = 80$ | | |
| **D.** | $72,000 \div 900 \longrightarrow 72 \div 9 = 8 \longrightarrow 72,000 \div 900 = 80$ | | |
| **E.** | $3,900 \div 130 \longrightarrow 39 \div 13 = 3 \longrightarrow 3,900 \div 130 = 30$ | | |

Use the above ideas to find the answers mentally.  The * number is the sum of the correct answers.

| | | | | | |
|---|---|---|---|---|---|
| **1.** | $480 \div 6$ | $3,500 \div 7$ | $240 \div 20$ | $56,000 \div 700$ | * 672 |
| **2.** | $480 \div 12$ | $4,500 \div 15$ | $7,500 \div 25$ | $4,600 \div 230$ | * 660 |
| **3.** | $3,600 \div 18$ | $1,000 \div 25$ | $5,400 \div 20$ | $2,000 \div 40$ | * 560 |
| **4.** | $880 \div 22$ | $6,600 \div 220$ | $690 \div 30$ | $27,000 \div 300$ | * 183 |
| **5.** | $450 \div 90$ | $3,800 \div 190$ | $600 \div 12$ | $7,000 \div 35$ | * 275 |
| **6.** | $3,900 \div 30$ | $640 \div 80$ | $750 \div 25$ | $8,400 \div 210$ | * 208 |
| **7.** | $290 \div 2$ | $3,200 \div 160$ | $693 \div 3$ | $9,000 \div 450$ | * 416 |
| **8.** | $800 \div 40$ | $33,000 \div 110$ | $780 \div 39$ | $8,100 \div 900$ | * 349 |

*Mental Math and Estimation*

# Mental Math Message

Use your mental math skills to find the message.  Write the letters above the correct answers.

**E** = 60 x 4 + 50          **I** = 97 − 48

**O** = 385 + 60             **T** = $70^2$

**H** = 23 x 30              **S** = 4500 ÷ 150

**N** = 200 x 42             **R** = 6 x 50 − 73

**A** = $3^2$ x 60           **U** = $60^2$ ÷ 18

**Y** = 1290 − 310           **D** = 34 x 200

$$\underline{\hspace{1cm}}\ \underline{\hspace{1cm}} \qquad \underline{\hspace{1cm}}\ \underline{\hspace{1cm}}\ \underline{\hspace{1cm}}\ \underline{\hspace{1cm}}\ \underline{\hspace{1cm}} \qquad \overset{I}{\underline{\hspace{1cm}}}\ \underline{\hspace{1cm}}$$

6800   445          4900  690  290  30  290          49   8400

$$\underline{\hspace{1cm}}\ \underline{\hspace{1cm}}\ \underline{\hspace{1cm}}\ \underline{\hspace{1cm}} \qquad \underline{\hspace{1cm}}\ \underline{\hspace{1cm}}\ \underline{\hspace{1cm}}\ \underline{\hspace{1cm}}$$

980  445  200  227          690  290  540  6800

# Mental Math Challenge

Enter the START number into your calculator.  Then enter the given operation (+, −, x, or ÷ ) and the number you think will give you the next circled answer.  Try to go from START to END without making a mistake.  Then try to go from END to START.

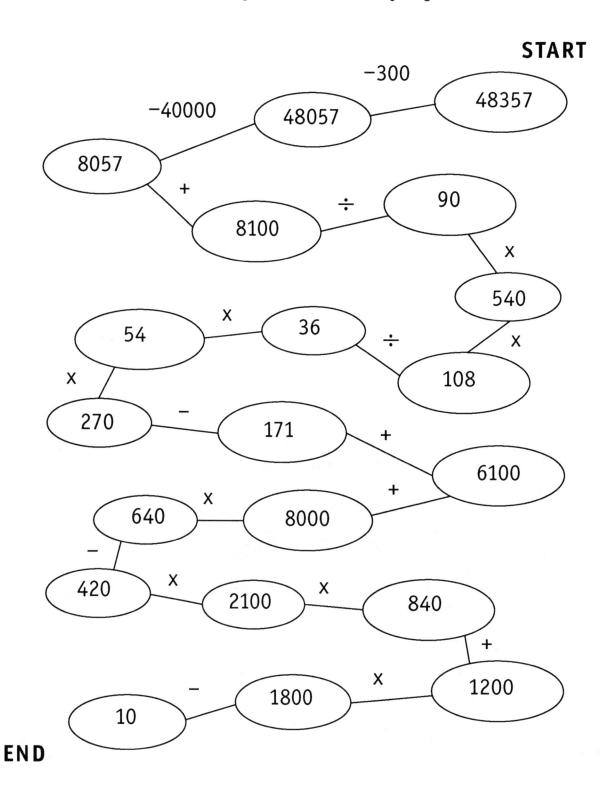

*Mental Math and Estimation* © 1993 Cuisenaire Company of America, Inc.

# Division Estimation

Looking at a *simpler* related problem is a way to estimate a quotient in a division problem.

A. $8\overline{)4796}$ ⟶ $8\overline{)4800}$ ⟶ 600

B. $4\overline{)193}$ ⟶ $4\overline{)200}$ ⟶ 50

C. $18\overline{)3472}$ ⟶ $18\overline{)3600}$ ⟶ 200

D. $7.8\overline{)547}$ ⟶ $8\overline{)560}$ ⟶ 70

E. $68\overline{)25718}$ ⟶ $70\overline{)28000}$ ⟶ 400

F. $0.47\overline{)246.8}$ ⟶ $5\overline{)2500}$ ⟶ 500

Try these — estimates only.

1. $6\overline{)3508}$ ⟶ _____

2. $29\overline{)8913}$ ⟶ _____

3. $91\overline{)7134}$ ⟶ _____

4. $4\overline{)19018}$ ⟶ _____

5. $6.9\overline{)4702}$ ⟶ _____

6. $13\overline{)3756}$ ⟶ _____

7. $38\overline{)2301}$ ⟶ _____

8. $5.8\overline{)416.7}$ ⟶ _____

9. $27\overline{)77131}$ ⟶ _____

10. $0.52\overline{)367.1}$ ⟶ _____

11. $36\overline{)697}$ ⟶ _____

12. $3.5\overline{)684.5}$ ⟶ _____

13. $58\overline{)2289}$ ⟶ _____

14. $0.47\overline{)3487}$ ⟶ _____

15. $37\overline{)2219}$ ⟶ _____

16. $14\overline{)4385}$ ⟶ _____

# Estimation Challenges

Use the completed examples and your estimation skills to find the missing numbers mentally. Record your mental answers, then check with a calculator. Score 1 point for each correct mental answer.

1.  4.2 x 55 = 231 ⟶ 0.42 x 550 = _____

2.  0.48 x 95 = 45.6 ⟶ 9.5 x 48 = _____

3.  6.2 x 5.5 = 34.1 ⟶ 3410 ÷ 62 = _____

4.  54.6 ÷ 8.4 = 6.5 ⟶ 840 x 0.65 = _____

5.  230 ÷ 0.92 = 250 ⟶ 2300 ÷ 92 = _____

6.  36 x 85 = 3060 ⟶ 306 ÷ 36 = _____

7.  6.8 x 57 = 387.6 ⟶ 570 x 0.068 = _____

8.  357 ÷ 0.42 = 850 ⟶ 3.57 ÷ 4.2 = _____

9.  31590 ÷ 70.2 = 450 ⟶ 7.02 x 450 = _____

10. 7.8 x 7.8 = 60.84 ⟶ 608.4 ÷ 78 = _____

11. 90.9 ÷ 1.8 = 50.5 ⟶ 5.05 x 180 = _____

12. 30 ÷ 0.48 = 62.5 ⟶ 4.8 x 62.5 = _____

My score _____

# Missing Number Challenges

Use your estimation skills and a calculator to find the missing whole numbers.

|     | M   | N   | M + N | M − N | M x N |
| --- | --- | --- | ----- | ----- | ----- |
| 1.  | ___ | ___ | 100   | ___   | 2356  |
| 2.  | ___ | ___ | 100   | 54    | ___   |
| 3.  | ___ | ___ | ___   | 58    | 2880  |
| 4.  | ___ | ___ | 98    | 46    | ___   |
| 5.  | 71  | ___ | ___   | 42    | ___   |
| 6.  | ___ | 14  | ___   | 73    | ___   |
| 7.  | ___ | ___ | 69    | ___   | 488   |
| 8.  | ___ | ___ | ___   | 58    | 603   |

# Estimation Challenges

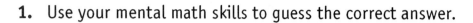

1.  Use your mental math skills to guess the correct answer.

    **a.** M + N = 100     M = 36     M x N = ?     1824     2604     2304     1234

    **b.** M + N = 100     M = 27     M x N = ?     2401     1971     1381     2731

    **c.** M + N = 100     M = 75     M x N = ?     1375     1525     2295     1875

    **d.** M + N = 100     M = 43     M x N = ?     2451     2191     2601     1891

2.  Use your mental math skills to guess the missing digits.

    **a.** M = 57   N = 43     77 x M = __38__     $(M - N)^2$ = __9__     89 x N = __82__

    **b.** M = 67   N = 83     M x N = __56__     $N^2$ = __88__     76 x N = __30__

    **c.** M = 49   N = 85     M x N = __16__     $2M^2$ = __80__     71 x N = __03__

3.  Use your estimation skills to place the decimal point in each answer. The first three nonzero digits are given.

    **a.** 7.395 x 5.713 ⟶ 422     **b.** 8.9 x 0.48 ⟶ 427

    **c.** 86.32 x 0.39 ⟶ 336     **d.** 26.8 x 0.32 ⟶ 857

    **e.** 756 x 0.24 ⟶ 181     **f.** 91.3 x 7.15 ⟶ 652

    **g.** 55.3 x 0.04 ⟶ 221     **h.** 31.9 x 2.6 ⟶ 829

4.  Use your estimation skills to decide whether each statement is true (reasonable) or false (not reasonable).

    **a.** 38 x 57 = 2166     **b.** 23 x 78 = 1564     **c.** 81 x 81 = 6561

    **d.** 97 x 46 = 4462     **e.** 17 x 68 = 1386     **f.** 48 x 45 = 2490

5.  Check your answers with a calculator.

# Mental Math Challenges

Use your mental math skills to find the missing numbers. Check with a calculator after recording your mental answers.

Example:

|  | x 45 | + 1230 | ÷ 50 | x 700 |
|---|---|---|---|---|
| 6 | 270 | 1500 | 30 | 21000 |

**1.**

|  | ÷ ___ | x ___ | − 270 | ÷ 7 |
|---|---|---|---|---|
| 3500 | 50 | 550 | | |

**2.**

|  | − 2.9 | + 3.58 | x 76 | ÷ 20 |
|---|---|---|---|---|
| 8.32 | | | | |

**3.**

|  | ÷ 20 | x 4 | + 734 | ÷ ___ |
|---|---|---|---|---|
| 380 | | | | 9 |

**4.**

|  | + 275 | x 0.09 | x 7 | − 90 |
|---|---|---|---|---|
| 325 | | | | |

**5.**

|  | ÷ ___ | x 1.5 | ÷ 50 | x ___ |
|---|---|---|---|---|
| 3600 | 300 | | | 2100 |

**6.**

|  | + ___ | ÷ ___ | x ___ | x 0.2 |
|---|---|---|---|---|
| 1700 | 2250 | 22.5 | 450 | |

© 1993 Cuisenaire Company of America, Inc.

# Mental Math Challenges

Use your mental math skills to find the missing numbers. Check with a calculator after recording your mental answers.

Example:

|  | x 24 |  | ÷ 16 |  | x 25 |  | ÷ 50 |  |
|---|---|---|---|---|---|---|---|---|
| 20 |  | 480 |  | 30 |  | 750 |  | 15 |

**1.**

|  | x ___ |  | ÷ ___ |  | ÷ ___ |  | x ___ |  |
|---|---|---|---|---|---|---|---|---|
| 32 |  | 9600 |  | 480 |  | 20 |  | 500 |

**2.**

|  | x ___ |  | ÷ ___ |  | x ___ |  | ÷ ___ |  |
|---|---|---|---|---|---|---|---|---|
| 25 |  | 1000 |  | 50 |  | 600 |  | 30 |

**3.**

|  | ÷ ___ |  | x ___ |  | ÷ ___ |  | x ___ |  |
|---|---|---|---|---|---|---|---|---|
| 3900 |  | 300 |  | 450 |  | 30 |  | 240 |

**4.**

|  | x ___ |  | ÷ ___ |  | x ___ |  | x 60 |  |
|---|---|---|---|---|---|---|---|---|
| 36 |  | 720 |  | 90 |  | 99 |  |  |

**5.**

|  | ÷ ___ |  | x ___ |  | ÷ ___ |  | x 6.5 |  |
|---|---|---|---|---|---|---|---|---|
| 440 |  | 40 |  | 480 |  | 30 |  |  |

**6.**

|  | ÷ ___ |  | x ___ |  | ÷ 20 |  | x 9 |  |
|---|---|---|---|---|---|---|---|---|
| 320 |  | 40 |  | 840 |  |  |  |  |

# Mental Math

Use your mental math skills to find the missing numbers. The * number is the sum of the correct answers.

Example:    3/8 of 72 ──► 3 x (72 ÷ 8) = 3 x 9 = 27

**1 a.** 2/3 of 24 = ———  **b.** 5/6 of 420 = ———  **c.** 1/2 of 3500 = ———

5/8 of 64 = ———  3/5 of 250 = ———  5/9 of 360 = ———

2/5 of 45 = ———  3/8 of 560 = ———  5/8 of 3200 = ———

3/4 of 88 = ———  2/3 of 480 = ———  3/4 of 2000 = ———

   * 140         * 1030         * 5450

Example:    5/6 of 420 ──► 5 x (420 ÷ 6) = 5 x 70 = 350

**2 a.** 2/3 of 660 = ———  **b.** 5/7 of 6300 = ———  **c.** 7/8 of 4000 = ———

3/8 of 640 = ———  4/5 of 3000 = ———  5/12 of 480 = ———

1/2 of 380 = ———  5/9 of 540 = ———  4/7 of 2800 = ———

3/4 of 200 = ———  7/25 of 750 = ———  4/9 of 8100 = ———

   * 1020         * 7410         * 8900

Example:    800 − 2/3 of 360 ──► 800 − 2 x (360 ÷ 3) = 800 − 240 = 560

**3 a.** 900 − 1/2 of 480 = ———  **b.** 1/2 of 680 + 260 = ———

2/3 of 900 − 390 = ———  450 + 3/4 of 120 = ———

3/5 of 250 − 60 = ———  400 − 5/6 of 240 = ———

3/4 of 320 + 270 = ———  2/3 of 990 − 250 = ———

   * 1470         *1750

# Mental Math

Use your mental math skills to find the missing numbers.  The number in each square must be the product of the two adjacent circled numbers.

Example:

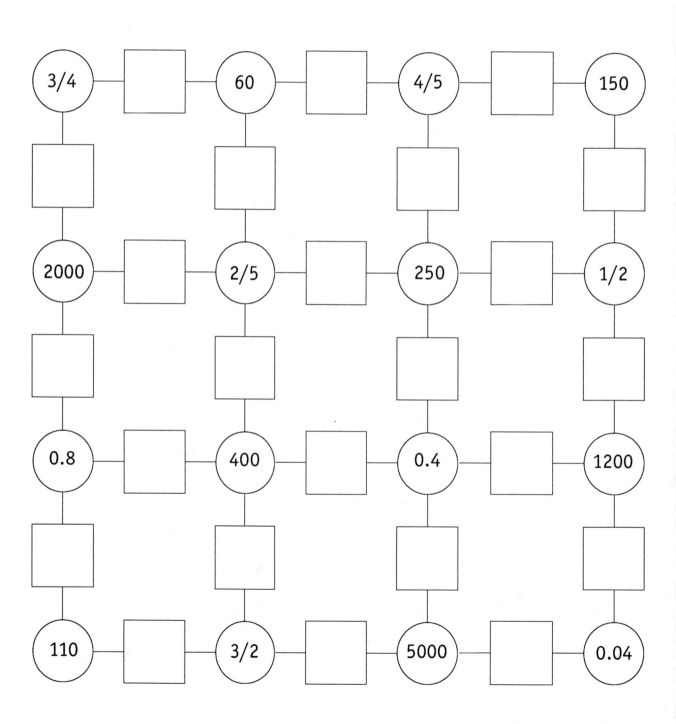

*Mental Math and Estimation*                                 © 1993 Cuisenaire Company of America, Inc.

# Mental Products

Use your mental math skills to find each answer.  The * number is the sum of the correct answers.

Example:     200 x 1.23 ⟶ 2 x (100 x 1.23) = 2 x 123 = 246

| | | | | |
|---|---|---|---|---|
| **1a.** 4000 x 0.08 | 600 x 0.9 | 70 x 0.7 | 400 x 0.12 | *957 |
| **b.** 0.15 x 200 | 200 x 3.4 | 2.2 x 30 | 20 x 4.5 | *866 |
| **c.** 6.3 x 200 | 300 x 0.13 | 500 x 0.2 | 300 x 2.5 | *2149 |
| **d.** 20 x 3.4 | 0.25 x 300 | 2.1 x 500 | 3.5 x 200 | *1893 |
| **e.** 600 x 0.8 | 2.4 x 200 | 20 x 3.9 | 400 x 1.5 | *1638 |

Example:     80 x 6 1/2 ⟶ 80 x (6 + 1/2) = 480 + 40 = 520

| | | | |
|---|---|---|---|
| **2a.** 60 x 4 1/2 | 200 x 7 1/2 | 400 x 6 1/4 | *4270 |
| **b.** 90 x 2 1/2 | 120 x 3 1/2 | 480 x 10 1/8 | *5505 |
| **c.** 200 x 2 1/4 | 220 x 4 1/2 | 360 x 2 1/3 | *2280 |
| **d.** 8 x 5.5 | 44 x 2.25 | 400 x 3.5 | *1543 |
| **e.** 6 x 9.5 | 80 x 4.25 | 240 x 2.25 | *937 |

# Mental Math Challenges

Use your mental math skills to find the missing numbers.  Check with a calculator after recording your mental answers.

Example:

|  | x 300 | + 260 | − 475 | + 625 |
|---|---|---|---|---|
| 2.3 | 690 | 950 | 475 | 1100 |

|  |  | x 8.5 | + 320 | − 280 | ÷ 19 |
|---|---|---|---|---|---|
| **1.** | 40 |  |  |  |  |

|  |  | x 800 | + 900 | − 1110 | ÷ 33 |
|---|---|---|---|---|---|
| **2.** | 1.5 |  |  |  |  |

|  |  | x 40 | + 560 | − 880 | ÷ 20 |
|---|---|---|---|---|---|
| **3.** | 21 |  |  |  |  |

|  |  | x 56 | + 192 | − 150 | ÷ 7 |
|---|---|---|---|---|---|
| **4.** | 8 |  |  |  |  |

|  |  | x 5 | + 480 | − 265 | ÷ 5 |
|---|---|---|---|---|---|
| **5.** | 64 |  |  |  |  |

|  |  | x 200 | + 370 | − 480 | ÷ 25 |
|---|---|---|---|---|---|
| **6.** | 1.8 |  |  |  |  |

*Mental Math and Estimation*

# Mental Math Challenges

Use your mental math skills to find the missing numbers.  Check with a calculator after recording your mental answers.

Example:

|  | x 0.7 | + 290 | − 80 | ÷ 9 |
|---|---|---|---|---|
| 600 | 420 | 710 | 630 | 70 |

**1.**

|  | x 80 | + 1900 | − 550 | ÷ 5 |
|---|---|---|---|---|
| 2.5 |  |  |  |  |

**2.**

|  | x 120 | + 1200 | − 2800 | ÷ 40 |
|---|---|---|---|---|
| 40 |  |  |  |  |

**3.**

|  | x 6 | + 256 | − 488 | ÷ 2 |
|---|---|---|---|---|
| 74 |  |  |  |  |

**4.**

|  | x 87 | + 211 | − 260 | ÷ 70 |
|---|---|---|---|---|
| 7 |  |  |  |  |

**5.**

|  | x 210 | + 79 | − 99 | x 5 |
|---|---|---|---|---|
| 0.4 |  |  |  |  |

**6.**

|  | x 200 | − 90 | ÷ 2 | + 495 |
|---|---|---|---|---|
| 1.9 |  |  |  |  |

# Mental Math Challenges

Use your mental math skills to find the missing numbers. Check with a calculator after recording your mental answers.

Example:

| | + 26 | x 80 | ÷ 36 | − 121 |
|---|---|---|---|---|
| 64 | 90 | 7200 | 200 | 79 |

| | | + | x | ÷ | − |
|---|---|---|---|---|---|
| 1. | 27 | 200 | 4800 | 300 | 152 |

| | | + | x | ÷ | − |
|---|---|---|---|---|---|
| 2. | 240 | 390 | 7800 | 200 | 137 |

| | | + | x | ÷ | − |
|---|---|---|---|---|---|
| 3. | 69 | 250 | 1000 | 200 | 58 |

| | | + | x | ÷ | − |
|---|---|---|---|---|---|
| 4. | 37 | 120 | 480 | 80 | 74.6 |

| | | + | x | ÷ | − |
|---|---|---|---|---|---|
| 5. | 17.6 | 30 | 750 | 3 | 1.85 |

| | | + | x | ÷ | − |
|---|---|---|---|---|---|
| 6. | 6.4 | 40 | 880 | 110 | 98.5 |

*Mental Math and Estimation*

# Mental Math Challenges

Use your mental math skills to find the missing numbers.  Check with a calculator after recording your mental answers.

Example:

|  | × 30 | + 431 | − 325 | ÷ 25 |
|---|---|---|---|---|
| 2.3 | 69 | 500 | 175 | 7 |

|  | × | + | − | ÷ |
|---|---|---|---|---|
| **1.** 45 | 900 | 1280 | 800 | 40 |

|  | × | + | − | ÷ |
|---|---|---|---|---|
| **2.** 70 | 56 | 140 | 98 | 49 |

|  | × | + | − | ÷ |
|---|---|---|---|---|
| **3.** 0.9 | 540 | 800 | 575 | 115 |

|  | × | + | − | ÷ |
|---|---|---|---|---|
| **4.** 1.5 | 120 | 400 | 270 | 9 |

|  | × | + | − | ÷ |
|---|---|---|---|---|
| **5.** 24 | 480 | 860 | 690 | 30 |

|  | × | + | − | ÷ |
|---|---|---|---|---|
| **6.** 8000 | 4800 | 7000 | 3500 | 7 |

# Mental Percents

| | | |
|---|---|---|
| 1% of 500 = 5 | 1% of 25000 = 250 | 1% of 60 = 0.6 |
| 10% of 570 = 57 | 10% of 7600 = 760 | 10% of 93 = 9.3 |

4% of 12000 = 4 x (1% of 12000) = 4 x 120 = 480

30% of 2200 = 3 x (10% of 2200) = 3 x 220 = 660

Use the above ideas to find the answer mentally. The * number is the sum of the correct answers.

1. 20% of 300       6% of 8000       8% of 400       * 572

2. 6% of 700       40% of 120       80% of 90       * 162

3. 70% of 11000       9% of 800       60% of 5000       * 10772

4. 6% of 450       8% of 7000       30% of 310       * 680

5. 30% of 1500       7% of 1200       80% of 600       * 1014

6. 50% of 5000       3% of 3300       2% of 280       * 2604.6

*Mental Math and Estimation*

# Mental Percents

Use your mental math skills to find the answers. The * number is the sum of the correct answers.

Example:     6% of 250 = 6 x (1% of 250) = 6 x 2.5 = 12 + 3 = 15

| | | | |
|---|---|---|---|
| **1a.** 3% of 900 | 7% of 500 | 9% of 2000 | * 242 |
| **b.** 6% of 400 | 2% of 1200 | 8% of 300 | * 72 |
| **c.** 5% of 6000 | 7% of 400 | 3% of 1300 | * 367 |

Example:     30% of 210 = 3 x (10% of 210) = 3 x 21 = 63

| | | | |
|---|---|---|---|
| **2a.** 20% of 230 | 30% of 1500 | 40% of 45 | * 514 |
| **b.** 80% of 250 | 30% of 800 | 20% of 170 | * 474 |
| **c.** 70% of 70 | 40% of 130 | 30% of 2200 | * 761 |

Examples:     4% of 2200 = 4 x 22 = 88

                   40% of 2200 = 4 x 220 = 880

| | | | |
|---|---|---|---|
| **3a.** 6% of 700 | 40% of 120 | 80% of 90 | * 162 |
| **b.** 30% of 1500 | 7% of 1200 | 80% of 600 | * 1014 |
| **c.** 6% of 450 | 8% of 7000 | 30% of 310 | * 680 |

# Estimating Percents

| | | |
|---|---|---|
| 1% = 1/100 = 0.01 | 6% = 6 x 1% = 0.06 | 8% = 8 x 1% = 0.08 |
| 1% of 700 = 700/100 = 7 | 8% of 700 = 8 x (1% of 700) = 8 x 7 = 56 | |

| | |
|---|---|
| 10% = 10/100 = 1/10 = 0.1 | 80% = 8 x 10% = 8 x 0.1 = 0.8 |
| 10% of 220 = 220/10 = 22 | 30% of 220 = 3 x (10% of 220) = 3 x 22 = 66 |

| | |
|---|---|
| 22% of \$278 ⟶ 20% of \$300 = \$60 | 7% of 219 ⟶ 7% of 200 = 14 |
| 39% of 613 ⟶ 40% of 600 = 240 | 8% of 486 ⟶ 8% of 500 = 40 |
| 53% of 1784 ⟶ 50% of 1800 = 900 | 6% of 79 ⟶ 6% of 80 = 4.8 |

Use the above ideas to estimate the answers mentally.

**1.** 29% of \$495 ⟶ _____ ⟶ _____

**2.** 17% of 4238 ⟶ _____ ⟶ _____

**3.** 6% of \$912 ⟶ _____ ⟶ _____

**4.** 4% of 2197 ⟶ _____ ⟶ _____

**5.** 6% of \$888

**6.** 28% of 413

**7.** 71% of 786

**8.** 87% of 72

**9.** 4% of 2478

**10.** 8% of \$689

**11.** 3% of 1476

**12.** 2% of 1766

**13.** 12% of 3847

*Mental Math and Estimation*

# Estimating Percents

1. Use your estimation skills to find the missing numbers from the list of possible answers.

   **a.** 31 is 25% of _____
   **b.** 15% of _____ is 102

   **c.** 154 is 35% of _____
   **d.** 12% of _____ is 96

   **e.** 33 is 6% of _____
   **f.** 5% of _____ is 11

   | Possible Answers | |
   | --- | --- |
   | 800 | 124 |
   | 990 | 220 |
   | 550 | 350 |
   | 440 | 680 |

2. Use your mental math skills to decide whether each statement below is true (reasonable) or false (not reasonable). Check with a calculator.

   **a.** 35% of 360 = 126
   **b.** 25% of 9600 = 2850

   **c.** 4% of 360 = 10.4
   **d.** 42% of 85 = 30.7

   **e.** 15% of 480 = 72
   **f.** 8% of 390 = 24.9

   **g.** 62% of 350 = 296.5
   **h.** 85% of 888 = 754.8

   **i.** 24% of 350 = 84
   **j.** 1.5% of 9600 = 144

   **k.** 0.5% of 9600 = 480
   **l.** 6% of 185 = 21.1

# Estimating Percents

Estimating an answer to a percent problem is often made easier by replacing the percent with a simple fraction.

| | | | |
|---|---|---|---|
| 10% = 1/10 | 20% = 1/5 | 25% = 1/4 | 33 1/3% = 1/3 |
| 40% = 2/5 | 50% = 1/2 | 75% = 3/4 | 66 2/3% = 2/3 |

48% of $179 ⟶ 1/2 of $180          26% of 1534 ⟶ 1/4 of 1600

35% of 3476 ⟶ 1/3 of 3600          65% of $38 ⟶ 2/3 of $390

9.2% of 498 ⟶ 1/10 of 500          77% of 2359 ⟶ 3/4 of 2400

Use the ideas above to *estimate* each product mentally.

**1.** 52% of $59.95 ⟶ _____ ⟶ _____

**2.** 24% of 2543 ⟶ _____ ⟶ _____

**3.** 32% of $43.95 ⟶ _____ ⟶ _____

**4.** 73% of 2936 ⟶ _____ ⟶ _____

**5.** 23% of 1729          **6.** 35% of $261          **7.** 18% of $43.75

**8.** 68% of 237          **9.** 53% of 6511          **10.** 12% of $478

**11.** 77% of 1467          **12.** 41% of $387          **13.** 27% of $469

# Estimating Percents

Use your estimation skills to guess the *best* answer for each of the problems.
The * number is the sum of the best answers.

**1 a.**  22% of 650 = N      N = ?   172      94      143      158

  **b.**  37 = N% of 185      N = ?   12      16      20      25

  **c.**  N% of 84 = 21      N = ?   8      15      18      25

  **d.**  34% of 1750 = N      N = ?   815      385      595      705

  **e.**  140% of 355 = N      N = ?   497      584      398      321

                                              *1280

**2 a.**  24% of 1750 = N      N = ?   420      340      590      700

  **b.**  57 = N% of 95      N = ?   60      42      51      70

  **c.**  18% of 5500 = N      N = ?   990      1530      680      1740

  **d.**  N% of 750 = 60      N = ?   12      4      8      6

  **e.**  N% of 1850 = 777      N = ?   31      42      26      51

                                              *1520

**3 a.**  35% of 160 = N      N = ?   5.6      56      4.8      48

  **b.**  195% of 140 = N      N = ?   27.3      273      22.6      226

  **c.**  36 = N% of 48      N = ?   35      45      55      75

  **d.**  7.5% of N = 12      N = ?   160      84      112      55

  **e.**  37.5% of N = 27      N = ?   58      72      98      124

                                              *636

# Using a Simpler Problem

Estimating an answer to a problem is often made easier way to by looking at a simpler related problem.

| | | | |
|---|---|---|---|
| 1/2 of $47.29 | → 1/2 of $48 | → $24 | |
| 2/3 of $382 | → 2/3 of $390 | → $260 | |
| 3/5 of 4601 | → 3/5 of 4500 | → 2700 | |
| 5/12 of 4478 | → 1/2 of 4000 | → 2000 | |
| 48% of $629 | → 1/2 of $600 | → $300 | |
| 35% of $888 | → 1/3 of $900 | → $300 | |

Use the above idea to *estimate* each product mentally.

**1.** 3/4 of 317    **2.** 5/8 of 713    **3.** 2/3 of 1746

**4.** 4/5 of $34    **5.** 4/9 of 537    **6.** 3/8 of 713

**7.** 24% of $158    **8.** 4.9% of 6187    **9.** 47% of 932

**10.** 5/6 of 4754    **11.** 18.9% of 479    **12.** 35% of $85

*Mental Math and Estimation*

# Percent Estimation Game

1. This is a game for 3 to 5 players.

2. One players selects a playing board, picks a problem on that board, and computes the answer with a calculator. The player then tells the others the board number and the problem he or she selected.

3. The player whose guess is closest to the correct answer scores 1 point.

4. The first player to score 5 points wins the game.

---

**Playing Board 1**

| | | |
|---|---|---|
| **a.** 47% of $63.95 | **b.** 24% of $34.99 | **c.** 38% of $77.95 |
| **d.** 66% of $31.98 | **e.** 15% of $79.88 | **f.** 74% of $41.99 |
| **g.** 6.9% of 87.8 | **h.** 2.5% of 3970 | **i.** 8.9% of 426 |
| **j.** 1.75% of 408 | **k.** 5% of 3575 | **l.** 4.5% of 7998 |

---

**Playing Board 2**

| | | |
|---|---|---|
| **a.** 149% of 698 | **b.** 0.49% of 6550 | **c.** 73% of 4768 |
| **d.** 7.8% of 312 | **e.** 9.5% of 52.8 | **f.** 1.9% of 879 |
| **g.** 39% of 1497 | **h.** 19.8% of 6139 | **i.** 73% of 4768 |

---

**Playing Board 3**

| | |
|---|---|
| **a.** 4000 − 39% of 7108 | **b.** 8% more than 7985 |
| **c.** 4.5% of 2995 − 135 | **d.** 2985 + 15% of 2985 |
| **e.** 89 − 33% of 89 | **f.** 8979 + 6% of 8979 |
| **g.** 39% of 719 − 99 | **h.** 24% less than 2457 |

# Mental Math Challenges

Use your mental math skills to find each answer.  The * number is the sum of the correct answers.

1. 36 + 49          67 + 78          45 + 58          * 333

2. 28 + 37 + 46     17 + 47 + 39     67 + 24 + 38     * 343

3. 92 − 47          71 − 29          183 − 38         * 232

4. 7 x 800          40 x 50          900 x 4          * 11200

5. 450 ÷ 9          2800 ÷ 40        3900 ÷ 130       * 150

6. 8 x 70 − 90      800 − 6 x 70     1000 − 3 x 90    * 1580

7. 600 x 0.4        0.3 x 220        1.5 x 460        * 996

8. 3/4 x 800        5/8 x 48         2/3 x 480        * 950

9. 6% of 500        5% of 80         3% of 2500       * 109

10. 40% of 60       30% of 120       15% of 400       * 120

# Mental Math Challenges

Use your mental math skills to solve each problem. The * number is the sum of the correct answers.

**1 a.** 20% of 4500 decreased by 350 is A.  A = _____

 **b.** 160 increased by 3/4 of 320 is B.  B = _____

 **c.** 290 more than 30% of 700 is C.  C = _____

 **d.** 1500 less than 5/8 of 7200 is D.  D = _____

 **e.** 800 decreased by 2/3 of 390 is E.  E = _____

 * 4990

**2 a.** The average of 80, 95, and 65 is A.  A = _____

 **b.** 5000 decreased by 2/3 of 4500 is B.  B = _____

 **c.** There are C hours in 5 days.  C = _____

 **d.** 3/4 of 240 is D less than 400.  D = _____

 **e.** 5/6 of 420 exceeds 99 by E.  E = _____

 * 2671

**3 a.** At 1/3 off, a $450-TV set will cost A.  A = _____

 **b.** 40% more than $60 is B.  B = _____

 **c.** 20% less than 1/2 of $1200 is C.  C = _____

 **d.** 3/8 of $4800 decreased by $850 is D.  D = _____

 **e.** At 25% off, a $160-bike will cost E.  E = _____

 * $1934

# Estimation Challenges

1. Use your estimation skills to place the decimal point in each answer. The first three nonzero digits are given.

   a. 140 x 0.49 ⟶ 6 8 6

   b. 250 x 0.075 ⟶ 1 8 7

   c. 26.88 ÷ 2.4 ⟶ 1 1 2

   d. 16.3 ÷ 0.95 ⟶ 1 7 1

   e. 12.167 ÷ 23 ⟶ 5 2 9

   f. 6.4% of 888 ⟶ 5 6 8

   g. 38.9% of 5477 ⟶ 2 1 3

   h. 1.87% of 1240 ⟶ 2 3 1

2. Use your estimation skills to find the missing numbers from the list of Possible Answers.

   a. _____ = 6.72 x 45

   b. _____ = 30.24 ÷ 6.72

   c. _____ = 3.024 ÷ 0.45

   d. _____ = 0.672 x 4500

   e. _____ = 302.4 ÷ 4.5

   f. _____ = 6.72 x 0.45

   g. _____ = 3024 ÷ 67.2

   h. _____ = 30.24 ÷ 45

   | Possible Answers | |
   | --- | --- |
   | 3024 | 0.672 |
   | 6.72 | 3.024 |
   | 30.24 | 67.2 |
   | 302.4 | 45 |
   | 4.5 | 672 |

3. Use your mental math skills to find the missing digits that will make each sentence true.

   a. 37 x (61 − 29) = __18__

   b. 43 x (72 − 37) = __50__

   c. 87 x (83 − 28) = __78__

   d. 96 x (63 − 36) = __59__

   e. __7 x 78 = 3666

   f. __4 x 69 = 5796

   g. __8 x 68 = 3264

   h. __9 x 53 = 4187

*Mental Math and Estimation*

# Solving Equations Mentally

**A.** $3N + 400 = 520 \longrightarrow 3N = 520 - 400 = 120 \longrightarrow N = 40$

**B.** $3N - 25 = 65 \longrightarrow 3N = 65 + 25 = 90 \longrightarrow N = 30$

**C.** $30N + 90 = 300 \longrightarrow 30N = 300 - 90 = 210 \longrightarrow N = 7$

Use the above idea to find the answers mentally. The * number is the sum of the correct answers.

**1.** $2N + 70 = 190$     $80N - 80 = 480$     $5N - 450 = 4050$     *967

**2.** $6N - 230 = 190$     $60N + 500 = 2900$     $30N + 800 = 3200$     *190

**3.** $70N - 650 = 2850$     $7N - 900 = 5400$     $4N + 80 = 360$     *1020

**4.** $2N + 70 = 190$     $5N - 40 = 160$     $7N - 1500 = 2700$     *700

**5.** $300N + 800 = 3200$     $5N + 450 = 1000$     $30N - 900 = 900$     *178

**6.** $40N + 900 = 1220$     $5N + 800 = 4300$     $4N + 90 = 410$     *788

**7.** $6N - 50 = 370$     $50N + 70 = 420$     $4N - 480 = 320$     *277

# Solving Equations Mentally

Use your mental math skills to solve each equation. The * number is the sum of the correct answers.

Example:     $3N + 90 = 300 \longrightarrow 3N = 300 - 90 = 210 \longrightarrow N = 70$

**1 a.** $2N + 70 = 190$      $10N - 40 = 240$      $60N + 500 = 2900$    * 128

   **b.** $70N - 650 = 2850$      $7N - 1500 = 2700$      $80N - 80 = 480$    * 657

   **c.** $30N + 800 = 3200$      $800N - 800 = 4000$      $15N - 450 = 4050$    * 386

Example:     $30N - 800 = 1300 \longrightarrow 30N = 2100 \longrightarrow N = 70$

**2 a.** $30N - 800 = 1900$      $4N + 90 = 410$      $50N + 800 = 4300$    * 240

   **b.** $20N + 900 = 2300$      $90N - 90 = 450$      $7N + 120 = 400$    * 116

   **c.** $6N - 230 = 190$      $40N + 900 = 1220$      $7N + 650 = 5550$    * 778

Example:     $3N + 70 = 250 \longrightarrow 3N = 250 - 70 = 180 \longrightarrow N = 60$

**3 a.** $6N - 230 = 190$      $60N + 500 = 2900$      $30N + 800 = 3200$    * 190

   **b.** $2N + 70 = 190$      $80N + 70 = 630$      $5N - 450 = 4050$    * 967

   **c.** $70N - 650 = 2850$      $3N + 90 = 360$      $7N - 900 = 5400$    * 1040

*Mental Math and Estimation*

# Equation Estimations

Use your mental math skills to guess the *best* answer for each problem.
The * number is the sum of the best answers.

**1 a.** $7N + 87 = 269$      $N = ?$     16   36   46   26

   **b.** $9N - 306 = 181$      $N = ?$     83   73   63   53

   **c.** $5N + 305 = 700$      $N = ?$     69   99   89   79

   **d.** $4N + 185 = 337$      $N = ?$     38   48   58   68

   **e.** $6N - 285 = 171$      $N = ?$     66   56   86   76

   **f.** $42N - 795 = 423$      $N = ?$     19   49   29   49

                                                 * 301

**2 a.** $69N - 1850 = 423$      $N = ?$     60   70   30   50

   **b.** $56N + 1611 = 3963$      $N = ?$     32   42   52   62

   **c.** $27N + 419 = 2174$      $N = ?$     65   75   85   95

   **d.** $82N - 655 = 2461$      $N = ?$     28   38   48   58

   **e.** $76N - 1986 = 2118$      $N = ?$     44   54   64   74

   **f.** $43N + 986 = 2190$      $N = ?$     18   28   38   48

                                                 * 257

# Equation Estimations

The **bold** number in each problem is the *best* estimate for the correct answer. Explain why.

1. 73 x 54 = ?                      3400  **3900**  4400   4800

2. 632 + 217 – 388 = ?              350   **460**   580    1200

3. 674 – 243 + 485 = ?             700    800     **900**  1100

4. 2/3 of $46.88 = ?                $28   **$32**  $38     $42

5. 39 x 82 = ?                      2600   2800    **3200**  3600

6. 5/18 of $3729 = ?               $750   **$1000** $1400  $1800

7. 22.8% of 635 = ?                 100   **150**   200    250

8. 37 = N% of 185      N = ?        6      12      **20**   28

9. 786 more than 2/3 of 702 = ?     700    900     **1200** 1400

10. 152% of 488 = ?                 600   **750**   900    1200

11. 4 X + 185 = 337     X = ?       20     30      **40**   50

12. 8 X – 256 = 200     X = ?       50    **60**    70     80

13. 39% of 2086 = ?                 70     500     **800**  6000

14. 649 more than half of 714 = ?   800   **1000**  1200   1400

*Mental Math and Estimation*

# Estimation—True or False?

Use your estimation skills to decide whether each statement below is true (reasonable) or false (not reasonable). Check with a calculator. Make each false statement true by changing the **bold** number.

1. 8031 exceeds **3978** by 5953.

2. Three times **279** is 496 more than 497.

3. 285 is 417 less than **702**.

4. Twice 489 plus **247** is 1225.

5. At $4.85 per hour, you would earn **$194** in 5 days working 8 hours per day.

6. If your car averages 28 miles per gallon, you will use 44 gallons of gasoline on a **982**-mile trip.

7. If you buy 3 T-shirts at $4.95 each and a cassette tape for $3.00, your change for a $20 bill will be **$3.15**.

8. The number halfway between 382 and **738** is 560.

9. Half of 788 is **250** more than half of 488.

10. If you put $85 into your savings account every other month, your savings during a three-year period will amount to **$1530**.

11. At $1.84 per pound, 3.5 pounds of peanuts will cost **$5.04**.

12. A square with a perimeter **76** cm has an area 361 cm$^2$.

# Estimation—True or False?

Use your estimation skills to decide whether each statement below is true (reasonable) or false (not reasonable). Check with a calculator. Make each false statement true by changing the **bold** number.

1.   Averaging **46** mph, it will take 13 hours to travel 598 miles.

2.   If Bill works 28 hours a week at **$4.25** per hour, his weekly earnings are $145.60.

3.   At $85 per month, Janet will save **$1600** in two years.

4.   A square whose area is 441 square inches has a perimeter of **84** inches.

5.   Typing at a rate of 55 words per minute, Jill will type **5600** words in two hours.

6.   A jet plane traveling at a speed of **650** mph will travel 2475 miles in 4.5 hours.

7.   If Kathy spends 20% of her $30,000 yearly income to rent an apartment, then her rent per month is **$600**.

8.   A bicycle priced at $180 will cost **$135** at a "25% off" sale.

9.   There are **504** hours in three weeks.

# Missing Digit Estimates

Use your mental math skills to guess the missing digit that will make each statement true.  Check with a calculator.  Score 1 point for each correct guess.

1.  There are ___72 hours in four weeks.

2.  169 less than half of 1032 is ___ 47.

3.  Half of ___ 98 divided by 19 is 21.

4.  48% of 850 increased by ___ 82 is 990.

5.  If 12 balls cost $8.16, then one ball will cost ___8 cents.

6.  At a rate of 3 for $1.29, one dozen oranges will cost $___.16.

7.  At 58 miles per hour, you can travel ___35 miles in 7.5 hours.

8.  There are ___44 hours during the month of December.

9.  A $92.00-radio will sell for $___9.00 at a "25% off" sale.

10. A rectangle having a width of 46 cm and a length of ___2 cm has an area of 3772 cm².

11. A square whose perimeter is 316 cm has an area of ___241 cm².

12. A cube with an edge of 31 cm has a surface area of ___766 cm².

13. If 5 bricks weigh 19.5 pounds, 80 bricks will weigh ___ 12 pounds.

14. If M − N = 10 and M = 72, then the product of M and N is ___464.

My Score _____

# Best Estimate

Use your mental math skills to guess the *best* answer.  Check with a calculator.
Score 1 point for each correct estimate.

1.  At 28% off, a $65-coat will cost about :    $15     $25     $35     $45

2.  A 6% tax on two $18-books is about:    $0.75    $1.30    $2.00    $2.90

3.  At 33% off, six $4.99-shirts will cost about:    $12    $16    $20    $25

4.  A 15% tip for a $6.95 meal will be about:    $0.50    $1.00    $1.50    $2.00

5.  One yard of rope costs 79 cents.  Nine pieces of rope, each 3 yards long, will cost
    about:                             $16    $22    $28    $34

6.  Working 28 hours per week at $6.75 an hour, Gina's salary for two months will
    be about:                       $800    $1000    $1200    $1600

7.  At $27,892 per year, Juanita's salary per month is about:

    $1,900    $2,300    $2,600    $2,900

8.  About 7.8% of the 76,278 people that live in Valley City are over 75 years old.  The
    number of people in Valley City that are over 75 years old is about:

    3,500    4,500    6,000    8,000

9.  About half of the 4,864 baseball fans at a game bought a hot dog for 75 cents.
    The total amount spent on hot dogs that day was about:

    $1,800    $2,600    $3,000    $3,700

10.  Allison exercises 15 minutes every day.  The number of hours she exercises in one
    year is about:                     60        90        120        1530

My Score _____

# Test Yourself (1)

| | |
|---|---|
| 47 + 95 = 142 | 28 + 57 + 63 = 148 |
| 348 + 460 = 808 | 546 + 739 = 1285 |

Find the exact answer mentally.

1. 56 + 38

2. 78 + 64

3. 75 + 87 + 125

4. 29 + 76 + 58

5. 760 + 187

6. 467 + 299

7. 1380 + 490

8. 365 + 48 + 67

9. 260 + 547

10. 278 + 534

11. 4370 + 630

12. 369 + 475

# Test Yourself (2)

---

$$482 - 99 = 383 \qquad\qquad 81 - 28 = 53$$

$$731 - 297 = 434 \qquad\qquad 522 - 88 = 434$$

---

Find the exact answer mentally.

**1.** $84 - 39$

**2.** $73 - 48$

**3.** $672 - 59$

**4.** $703 - 96$

**5.** $712 - 299$

**6.** $888 - 189$

**7.** $601 - 198$

**8.** $183 - 47$

**9.** $800 - 675$

**10.** $813 - 298$

**11.** $732 - 389$

**12.** $852 - 380$

*Mental Math and Estimation*

# Test Yourself (3)

$5.00 − $1.85 = $3.15          $20.00 − $12.90 = $7.10

$1.00 − $0.49 = $0.51          $10.00 − $6.99 = $3.01

Find the exact answer mentally.

1.  $10.00 − $3.99

2.  $5.00 − $2.95

3.  $20.00 − $16.99

4.  $1.00 − $0.65

5.  $50.00 − $38.00

6.  $10.00 − $6.19

7.  $5.00 − $3.88

8.  $20.00 − $7.90

9.  $10.00 − $2.60

10. $20.00 − $13.75

11. $5.00 − $3.49

12. $50.00 − $19.50

# Test Yourself (4)

| | |
|---|---|
| 90 x 60 = 5400 | 25 x 30 = 750 |
| 3200 ÷ 80 = 40 | 4800 ÷ 160 = 30 |

Find the exact answer mentally.

**1.** 60 x 70

**2.** 12 x 40

**3.** 6300 ÷ 70

**4.** 39000 ÷ 130

**5.** 8 x 500

**6.** 20 x 360

**7.** 8600 ÷ 43

**8.** 6000 ÷ 200

**9.** 60 x 110

**10.** 8400 ÷ 210

**11.** 200 x 55

**12.** 750 ÷ 25

*Mental Math and Estimation*

# Test Yourself (5)

---

3/4 of 2400 = 3 x 2400/4 = 3 x 600 = 1800

2/3 of 660 − 90 = 2 x 220 − 90 = 440 − 90 = 350

---

Find the exact answer mentally.

1. 5/8 of 24

2. 2/3 of 660

3. 4/5 of 300

4. 5/12 of 240

5. 4/5 of 4500

6. 3/8 of 720

7. 2/3 of 600 + 189

8. 700 − 5/6 of 420

9. 1/2 of 7080

10. 900 − 1/2 of 640

11. 450 + 3/4 of 120

12. 2/3 of 990 − 250

---

# Test Yourself (6)

$$800 \times 0.6 = 80 \times 6 = 480 \qquad 0.035 \times 200 = 7$$

$$48 \div 2.4 = 480 \div 24 = 20 \qquad 35 \div 0.5 = 350 \div 5 = 70$$

Find the exact answer mentally.

1. $0.76 \times 1000$

2. $760 \div 100$

3. $0.05 \times 400$

4. $4800 \div 200$

5. $2.5 \times 6000$

6. $0.025 \times 400$

7. $55 \div 0.1$

8. $800 \times 0.25$

9. $480 \div 1.6$

10. $32 \div 0.4$

11. $2.25 \times 80$

12. $6000 \times 0.05$

*Mental Math and Estimation*

# Test Yourself (7)

---

1% of 2500 = 25          10% of 580 = 58

8% of 250 = 20           20% of 240 = 48

---

Find the exact answer mentally.

1. 6% of 500

2. 80% of 70

3. 4% of 1200

4. 30% of 250

5. 40% of 1500

6. 2% of 440

7. 150% of 46

8. 6% of 5500

9. 50% of 780

10. 0.5% of 6800

11. 80% of 250

12. 2.5% of 2200

# Test Yourself (8)

Find the exact answer mentally.

1. 50 x 800

2. 6350 + 400

3. 78 + 56 + 32

4. 3/8 of 5600

5. 831 − 299

6. 8 x 54

7. 4800 ÷ 16

8. 5 x 67 x 20

9. $20.00 − $6.75

10. 7 − 2 3/8

11. 162 − 29

12. 40 x 23

13. 9 x 80 − 70

14. 20% of 340

15. $3.75 + $5.50

16. 5% of 800

17. 40 x $0.75

18. 825 − 350

19. 150% of 600

20. 18 x 35

# Test Yourself (9)

Find the exact answer mentally.

1. There are 20 nickels in $1.00. How many nickels are there in $3.50?

2. How many minutes is it from 10:30 am to 5 pm?

3. A cafe ordered 40 dozen buns. How many buns were ordered?

4. If one school bus holds 36 students, how many students can ride on 8 buses?

5. Julie bought 15 bags of peanuts at 25 cents each and sold them for 55 cents each. How much money did she make?

6. Ken was to be in school at 8:15 am. He didn't get there until 9:28 am. How many minutes late was he?

7. Jenny had a $20-bill. She bought two shirts at $6.99 each. How much money does she have now?

8. Beth has $16.70. Ruth has $11.69. How much more money does Beth have than Ruth?

9. If apples cost $3.60 a dozen, how much will 40 apples cost?

10. How many hours are there in 20 days?

# Test Yourself (10)

Find the exact answer mentally.

1. If candy bars are on sale at 2 for 90 cents, how many could you buy for $4.50?

2. Kathy has a $5 bill, 3 quarters, 2 dimes, and 6 nickels.  How much money does she have?

3. A $450-stereo is on sale at 1/3 off.  What is the sale price?

4. Golf balls are on sale at 3 for $1.80.  How much will 8 of these balls cost?

5. James made 9 of 12 shots during his last basketball game.  What percent did he miss?

6. Of the 2,500 people at a convention, 60% were women.  How many people at the convention were not women?

7. How many minutes are there in one day?

8. How many hours in the month of April?

9. A $1500-computer is on sale at 20% off.  What is the sale price?

10. At $6.50 an hour, how much would 40 hours cost?

*Mental Math and Estimation*

# Estimation Check  (1)

Use your mental math skills to guess the missing digits.

1.  68 + 175 + 39 + 316 = __9__

2.  5,639 + 4,206 − 1,867 = __97__

3.  767 + 512 + 321 − 896 = __0__

4.  9,101 − (3,302 + 808) = __99__

5.  1,866 + 389 + 2,031 + 4,928 = __21__

6.  639 + 48 + 3,061 + 487 + 793 = __02__

7.  87 x 64 + 726 = __29__

8.  38 x 121 − 736 = __86__

9.  74 x 93 − 2,888 = __99__

10.  29(67 + 49 + 23) = __03__

11.  289(102 − 78) = __93__

12.  $41^2$ − 789 = __9__

# Estimation Check (2)

Use your mental math skills to guess the missing digits.

1.  49 x __2 = 3038

2.  82 x __2 = 4264

3.  76 x __4 = 4104

4.  85 x __3 = 2805

5.  __7 x 43 = 2021

6.  68 x __7 = 1836

7.  47 x __9 = 3243

8.  67 x __3 = 2881

9.  29 x __4 = 1856

10.  __2 x 53 = 3816

11.  __8 x 58 = 3944

12.  __3 x 66 = 4158

*Mental Math and Estimation*

# Estimation Check (3)

Use your mental math skills to guess the missing digits.

1. $8 \times 29^2 =$ __ 72 __

2. $8 \times 73^2 =$ __ 263 __

3. $7 \times 29^2 =$ __ 88 __

4. $4 \times 32^2 =$ __ 09 __

5. $6 \times 58^2 =$ __ 018 __

6. $7 \times 34^2 =$ __ 09 __

7. $6 \times 38^2 =$ __ 66 __

8. $7 \times 73^2 =$ __ 730 __

9. $3 \times 55^2 =$ __ 07 __

# Estimation Check  (4)

Use your estimation skills to decide whether each of the following equations is true (reasonable) or false (not reasonable).

1.   38 x 57 = 2166

2.   23 x 78 = 1564

3.   712 ÷ 38 = 24

4.   1711 ÷ 29 = 62

5.   48 x 45 = 2490

6.   81 x 81 = 6561

7.   1894 ÷ 37 = 62

8.   3588 ÷ 46 = 78

9.   17 x 68 = 1386

10.  97 x 46 = 4462

11.  912 ÷ 16 = 57

12.  1458 ÷ 27 = 54

*Mental Math and Estimation*

# Estimation Check (5)

Use your estimation skills to decide whether each of the following equations is true (reasonable) or false (not reasonable).

1. 8400 x 0.35 = 3940

2. 0.32 x 450 = 184

3. 2800 x 0.15 = 420

4. 96 x 0.16 = 15.36

5. 0.42 x 760 = 372.2

6. 2800 x 0.26 = 728

7. 5500 x 0.08 = 440

8. 64 x 0.45 = 28.8

9. 0.06 x 8500 = 510

10. 0.75 x 6800 = 5100

11. 18.6 x 0.65 = 7.09

12. 250 x 0.18 = 45

# Estimation Check (6)

Use your estimation skills to decide whether each of the following equations is true (reasonable) or false (not reasonable).

1. 35% of 360 = 126

2. 42% of 85 = 30.7

3. 35% of 480 = 198

4. 25% of 9500 = 2850

5. 20% of 480 = 96

6. 15% of 360 = 48

7. 30% of 64 = 19.2

8. 5% of 6400 = 320

9. 44% of 85 = 42.4

*Mental Math and Estimation*

# Estimation Check (7)

Use your estimation skills to guess the *best* answer mentally.

**1.** $74N + 86 = 2824$  $N = ?$  27  37  47  57

**2.** $79N - 711 = 2765$  $N = ?$  34  44  54  64

**3.** $43N + 679 = 3089$  $N = ?$  46  56  66  76

**4.** $74N + 808 = 5100$  $N = ?$  38  48  58  68

**5.** $9000 - 79N = 1969$  $N = ?$  69  79  89  99

**6.** $77N - 2492 = 819$  $N = ?$  33  43  53  63

**7.** $83N + 9269 = 11842$  $N = ?$  21  31  41  51

**8.** $5819 - 29N = 3760$  $N = ?$  41  51  61  71

**9.** $N^2 - 798 = 2923$  $N = ?$  41  51  61  71

**10.** $7000 - N^2 = 3864$  $N = ?$  36  46  56  66

**11.** $N^2 + 764 = 2285$  $N = ?$  19  29  39  49

**12.** $8800 - N^2 = 3324$  $N = ?$  44  54  64  74

# Estimation Check (8)

Use your estimation skills to guess to *best* answer mentally.

1. $3.76 + $0.69 + $4.25 + $1.87 + $5.78 + $0.39 is about:
   a. $13　　　　b. $19　　　　c. $15　　　　d. $17

2. Which problem below will give the best estimate for 19.4 x 46.1?
   a. 20 x 50　　b. 20 x 45　　c. 20 x 40　　d. 10 x 50

3. 96.7 + 147.4 + 62.75 + 36.8 is closest to:
   a. 350　　　　b. 300　　　　c. 400　　　　d. 250

4. 3 5/6 + 2 1/5 + 1 7/8 + 4 2/3 + 2 3/5 is closest to:
   a. 15　　　　b. 18　　　　c. 13　　　　d. 17

5. Which of the following will give the best estimate for 27/55 x 639?
   a. 2/5 x 640　b. 3/5 x 640　c. 1/3 x 640　d. 1/2 x 640

6. 3/4 of 97 5/6 is closest to:
   a. 90　　　　b. 75　　　　c. 60　　　　d. 45

7. 84 x 34 is closest to:
   a. 2400　　　b. 2800　　　c. 3200　　　d. 3600

8. If 24.5 x 32 = 784, then 2.45 x 3.2 equals:
   a. 0.0784　　b. 78.4　　　c. 0.784　　　d. 7.84

9. At $2.39 per yard, the cost for 2 1/2 yards of material will be about:
   a. $4.00　　　b. $5.00　　　c. $6.00　　　d. $7.00

10. 379 divided by 0.47 is about:
    a. 800　　　b. 600　　　c. 1800　　　d. 200

11. 4/9 + 6/7 is closest to:
    a. 0.5　　　b. 1　　　　c. 2　　　　d. 2.5

12. 3.49 x 58.7 is closest to:
    a. 1500　　　b. 150　　　c. 250　　　d. 200

*Mental Math and Estimation*

# Estimation Test (9)

Use your estimation skills to guess the *best* answer mentally.

**1.** 5/8 x 713 is closest to:

    **a.** 200        **b.** 300        **c.** 400        **d.** 550

**2.** 42,165 + 34,108 + 38,210 + 32,987 + 47,566 + 45,865 is about:

    **a.** 200,000    **b.** 240,000    **c.** 300,000    **d.** 360,000

**3.** 0.72 − 0.009 is closest to:

    **a.** 0.07        **b.** 0.7        **c.** 0.6        **d.** 0.06

**4.** 9.72 + 3.17 + 0.94 + 2.09 is closest to:

    **a.** 24        **b.** 18        **c.** 14        **d.** 16

**5.** 563.7 ÷ 2.93 is closest to:

    **a.** 280        **b.** 190        **c.** 20        **d.** 130

**6.** 3.87 x 1.94 x 5.23 is closest to:

    **a.** 15        **b.** 40        **c.** 20        **d.** 30

**7.** A clock loses 29 seconds every day. About how many minutes will it lose in one year?

    **a.** 120 min.    **b.** 900 min.    **c.** 720 min.    **d.** 180 min.

**8.** Hamburger was $1.69 per pound. Steak was $2.79 per pound. Randi bought 2 pounds of each type of meat and gave the clerk a $20 bill. About how much change did she get?

    **a.** $9.00    **b.** $7.00    **c.** $11.00    **d.** $15.00

**9.** The 17 teachers at Lincoln school paid $55 to buy a gift for their secretary. About how much did each teacher contribute?

    **a.** $2.50    **b.** $3.25    **c.** $4.50    **d.** $3.90

**10.** Last year our basketball team made 72% of their 350 free throws. About how many free throws did they make?

    **a.** 250    **b.** 300    **c.** 200    **d.** 150

# Estimation Check (10)

Use your estimation skills to guess the *best* answer mentally.

1. Notebooks were $2.89 each. Pens were $1.39 each.  Kate bought 3 notebooks and 5 pens. Her total cost was about:
   a. $19.00  b. $11.00  c. $13.00  d. $16.00

2. Beth jogs 40 minutes every day.  At this rate, about how many hours does she jog in one year?
   a. 14,000  b. 400  c. 240  d. 1,600

3. Gary has a part-time job working 4 days a week from 11:00 am to 4:30 pm.  At $7.25 per hour, his weekly salary is about:
   a. $240  b. $300  c. $390  d. $160

4. Golf balls are on sale at 3 for $1.74.  At this rate, 2 dozen golf balls cost about:
   a. $10.00  b. $15.00  c. $20.00  d. $42.00

5. Two meters of rope cost 78 cents.  About how much does a 68-meter long rope of this type cost?
   a. $27.00  b. $23.00  c. $32.00  d. $19.00

6. 1.4% of the people that live in Valley City are over 85 years old.  About how many of the 28,500 people that live there are 85 years old?
   a. 3,000  b. 300  c. 400  d. 40

7. Cindy bought a used car for $3,888.  She paid for it in 24 equal monthly installments. About how much was each payment?
   a. $160  b. $120  c. $190  d. $280

8. About half of the 31,964 fans at a baseball game bought a hot dog for $1.25.  The income that day for hot dogs was about:
   a. $48,000  b. $34,000  c. $20,000  d. $15,000

9. My yearly salary is now $27,800.  A 4.8% pay raise will increase my monthly income by about:
   a. $50  b. $100  c. $180  d. $250

*Mental Math and Estimation*  © 1993 Cuisenaire Company of America, Inc.